Keli Vanessa Salvador Damin

Optimization of self-lubricating composites

Keli Vanessa Salvador Damin

Optimization of self-lubricating composites

Presentation of a potential technique for optimizing self-lubricating composites

ScienciaScripts

Imprint

Cover image: www.ingimage.com

This book is a translation from the original published under ISBN 978-3-330-77271-7.

Publisher:
Sciencia Scripts
is a trademark of
Dodo Books Indian Ocean Ltd. and OmniScriptum S.R.L publishing group

120 High Road, East Finchley, London, N2 9ED, United Kingdom
Str. Armeneasca 28/1, office 1, Chisinau MD-2012, Republic of Moldova, Europe
Managing Directors: Ieva Konstantinova, Victoria Ursu
info@omniscriptum.com

Printed at: see last page
ISBN: 978-620-8-64326-3

SUMMARY

I dedicate this work to my mother, for all the love and effort she put in to ensure that my goals were always achieved.

1 INTRODUCTION

With the current concern for the environment and sustainable development, special importance is being given to energy efficiency, both in production processes and in the application of products. This greater energy efficiency can be achieved through greater control of the tribological properties of materials, i.e. control of the phenomena related to friction, lubrication and wear. In the 1960s, the OECD (Organization for Economic Cooperation and Development), an international organization made up of 34 countries and based in Paris, commissioned a study to investigate this issue. The study, known as the "Jost Report", estimated that the state could save 515 million pounds (in 1965 values) if more attention was paid to tribology (JOST, 1990). This study aroused interest in other industrialized countries, such as Canada, Japan and the USA, which produced similar reports (KUBOTA, 1982; MOLGAARD, 1984). One example is the work published by Pinkus and Wilcock in 1977, which assumed that 11% of energy consumption in the areas of road transportation, power generation and industrial processes and machinery could be saved in the United States if better tribological practices were adopted (PINKUS et al., 1977).

Statistics from developed countries have shown that around 1 to 6% of GDP is spent on wear and tear. At the same time, it is estimated that 20% of these losses could be avoided by applying existing knowledge about wear, friction and lubrication (KLEIN et al., 2010).

In passenger cars, one third of fuel energy is used to overcome friction in the engine, tires, brakes and transmission. In 2009, 208 billion liters of fuel were used worldwide to overcome friction in passenger cars. If new technologies for reducing friction in vehicles were implemented, friction losses could be reduced by 18% in the short term (5-10 years) and 61% in the long term (15-25 years). This would amount to savings of 174 and 576 billion euros respectively (HOLMBERG; ANDERSSON; ERDEMIR, 2012).

Based on the above, the production of materials with better tribological properties provides a longer service life and a reduction in costs, either through fewer part changes or machine *setups*, as well as minimizing the production of polluting gases. In this context, one way of increasing the useful life of materials, and thus the energy efficiency of the process, is to control friction and minimize wear by producing self-lubricating components.

Self-lubricating components can be produced in two ways: by applying the solid lubricant in the form of films or layers; or by incorporating the solid lubricant, in the form of dispersed particles, into the entire volume of the component, in which case the powder metallurgy (PM) technique can be used. The study of self-lubricating composites is one of the major lines of current research in which several studies have been and are being carried out (RAMOS FILHO et al., 2014; PARUCKER et al., 2014; BINDER et al., 2011; HAMMES, et al., 2014; FURLAN et al., 2014; SCHROEDER et al., 2012; BINDER et al., 2008).

Commonly used self-lubricating mechanical components with a low coefficient of friction, such as sintered bronze and copper self-lubricating bushings, have found use in a variety of equipment, such as drills, printers, blenders, among others. However, it is well known that these components do not have high mechanical strength due to the high volumetric content (15% to 40%) of solid lubricant particles, which results in a low degree of continuity of the matrix phase, which is precisely the element responsible for the material's mechanical strength (HAMMES et al., 2014; BINDER, 2009; PARUCKER et al., 2014).

Thus, one way of obtaining products with high mechanical resistance is through the use of composites of ferrous matrices with alloying elements, which are capable of increasing the mechanical resistance of the matrix due to microstructural modification, such as the formation of different phases or precipitates (THÜMMLER; OBERACKER, 1993).

This improvement in performance, however, comes at a high cost because, in this case, dispersed alloying elements must be added to the total volume of the component while, in most cases, an increase in resistance is only desired on the surface, because in most engineering applications, parts with wear in the order of a few micrometers (5-10 μm) already lose their function and dimensional tolerance causing failure in the mechanical system in which they are installed.

In this sense, the search for processes that improve mechanical and tribological properties at a lower cost than using alloying elements in powder form has become attractive. This book first covers the basics of solid and plasma lubrication, then describes what self-lubricating composites are and what surface enrichment is and how it can be applied to optimize self-lubricating composites.

2 THEORETICAL BACKGROUND

This section will deal with topics that are relevant to the understanding and development of this text. Firstly, the issue of solid lubrication will be addressed, as well as powder metallurgy, so that the reader can be directed to the topic of self-lubricating composites, where some developments in this area will be presented. Finally, a brief explanation of plasma will be presented, followed by the topic of surface enrichment, which will discuss the main theme of the book.

2.1 SOLID LUBRICANT

In most tribological applications, liquid lubricants or greases are used to combat friction and wear; but when the service conditions become unfavorable or very severe, solid lubricants are the option (ERDEMIR, 2001; DONNET; ERDEMIR, 2004). Some examples are: the food and pharmaceutical industry, where hydrodynamic lubrication is undesirable due to possible contamination of products and the environment; in high-temperature assemblies, where oil degradation is imminent; in situations of contact with extreme pressure; in high vacuum and/or extremely low temperature environments, among others (DONNET; ERDEMIR, 2004; ERDEMIR, 2005).

Basically, the purpose of a solid lubricant is to provide dry lubrication to two surfaces that are interacting in relative motion. There are various types of solid lubricants on the market, and based on their physical, chemical, structural and mechanical properties, they can be classified into various subcategories (ERDEMIR, 2001). However, in simplified terms, they can be classified into two large groups: hard and soft solid lubricants. Lubricants in the first group are those with a hardness greater than 10 GPa. They have greater resistance to wear and a lower coefficient of friction than soft solid lubricants. Examples of hard solid lubricants include oxides (Al2O3, Cr2O3), carbides (TiC, WC) and carbon-based coatings (such as DLC - Diamond like carbon) (DONNET; ERDEMIR, 2004). Soft solid lubricants have a hardness of less than 10 GPa and can have a low coefficient of friction, but not always high wear resistance

(BINDER, 2009). Examples of soft lubricants are some polymers (PE, PTFE), soft metals (Pb, Ag) and lamellar solids (MoS_2, hBN, graphite) (DONNET; ERDEMIR, 2004).

The solid lubricant can be applied to the components of a tribological pair in the form of films or composite layers deposited or generated on their surface (PARUCKER, et al., 2014; BINDER, 2009). In the case of application as films, these can be processed by numerous methods from simple spreading of fine powders on the surface to complex vacuum deposition systems such as physical and chemical vapor deposition (PVD and CVD), ion beam assisted deposition and pulsed laser deposition (HOGMARK et al., 2001; PAULEAU; THIERY, 2004; ERDEMIR; ERYILMAZ; KIM, 2014).

Another method for producing carbon films is carbon nanostructures derived from carbides (MCNALLAN et al., 2005; WELZ; GOGOTSI; MCNALLAN, 2003; KORENBLIT; YUSHIN, 2014; GOGOTSI et al., 2001; MCNALLAN et al., 2005). The advantage of this method is that it does not require high vacuum or plasma forming systems, unlike plasma, PVD and CVD methods, which are used to synthesize diamond and DLC films. The procedure is carried out under atmospheric pressure in a furnace using only chlorine gas (Cl), at temperatures between 600 and 1100° C, using components made from carbides (such as WC or SiC) as the base material. In this technique, the chlorine gas reacts preferentially with the metal atoms of the carbide, forming a volatile product that is eliminated from the furnace. In this way, the carbon atoms of the carbide reorganize themselves on the surface of the substrate forming carbon-based nanostructures in the form of thin films (GOGOTSI et al., 2001). Figure 1 shows different nanostructures derived from carbides.

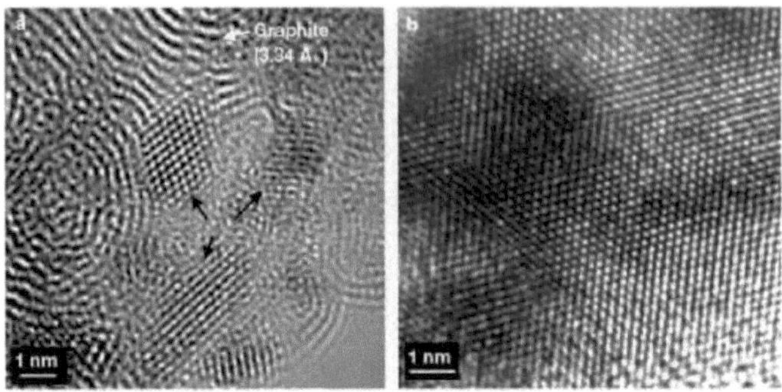

Figure 1 - TEM micrographs showing (a) diamond-structured carbon nanocrystals (indicated by the black arrows) (b) diamond-structured nanocrystalline carbon region (GOGOTSI et al., 2001).

Something to be aware of in the case of lubrication with coatings is that the durability of the lubricity regime is related to the thickness of the film, i.e. if the film is broken there will be metal-to-metal contact between the two surfaces and consequently damage to the unprotected surface of the material due to the unwanted tribological effects of high friction and wear (DONNET; ERDEMIR, 2004; ERDEMIR, 2001).

Another way of applying solid lubricant is by incorporating it into the volume of the component to be produced in the form of particles dispersed in its matrix (GERMAN, 1996; THÜMMIER; OBERACKER, 1993; DENG; CAN; SUN, 2005;ZALAZNIK et al., 2016; RODRIGUEZ et al., 2016; RAMOS FILHO, et al., 2014; PRABHU, 2015; MAHATHANABODEE et al., 2013). Figure 2 illustrates this situation.

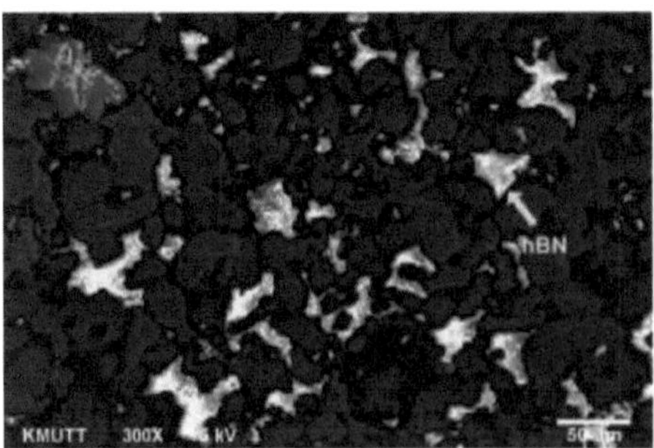

Figure 2 - Microstructure of a self-lubricating composite: 316L stainless steel with 20% vol. by volume of h-BN (MAHATHANABODEE et al., 2013).

This is the opposite of the situation with films, where the deposits of solid lubricant in the volume of the matrix allow for a constant supply of lubricant, preventing direct contact between the surfaces (DONNET; ERDEMIR, 2004; ERDEMIR, 2001). One of the ways of incorporating the solid lubricant into the volume of the material is through the powder metallurgy technique.

2.2 POWDER METALLURGY

Powder metallurgy (PM) is a process for manufacturing metal parts that differs from conventional metallurgical processes in that it uses both metallic and non-metallic powders as raw materials. Powder metallurgy is a manufacturing process whose use has been consolidated by the production of components with structural and physical characteristics that could hardly be obtained by any other manufacturing process (THÜMMLER; OBERACKER, 1993). Parts can be manufactured using powder metallurgy in several stages. But traditionally, the process is developed in four basic stages: powder production, powder mixing, compaction and sintering (THÜMMLER; OBERACKER, 1993).

One of the versatilities of the PM technique is the possibility of incorporating a solid lubricant into the volume of the matrix of the material to be manufactured and thus producing sintered self-lubricating composites (DE MELLO et al., 2011). These composite materials can be easily produced through powder metallurgy, a technique which has been widely used to produce parts in industry, mainly because of its low cost when applied to large volume production.

There are two ways of adding solid lubricant particles to the volume of a metal matrix using the PM technique (BINDER et al., 2008; BINDER et al., 2010):

1. Mixing solid lubricant powder with matrix metal powders by a simple mixing process; and
2. Generation of solid lubricant particles in situ during sintering by

dissociation of a precursor mixed with the metal matrix powders.

The advantage of generating solid lubricant particles in situ is that in this case the microstructure of the composite is optimized due to the high degree of continuity of the metal matrix. As in the first case, there is no formation of interposed layers of solid lubricant between the metal particles of the matrix, which reduces the mechanical strength of the composite (BINDER et al., 2008; BINDER, 2009; BINDER et al., 2010).

In both cases described above, PM is a very versatile technique in the development of self-lubricating composites because it makes it easier to study different process variables, either because it is easy to choose powders of different natures or because it is easy to use different chemical compositions, making it a very promising method in the development of sintered self-lubricating composites. Further details on the two techniques mentioned above will be discussed below, where results on the subject will be presented.

2.3 SELF-LUBRICATING coMPosiTS

The term composite refers to a material formed by combining at least two different materials, with a defined interface between them, combining complementary properties that would not be possible with the components alone. In this way, a self-lubricating composite can be generated by simply mixing a solid lubricant powder with a material matrix using the PM technique (DE MELLo et al., 2010).

self-lubricating composites are used to increase the service life of machinery and equipment where hydrodynamic lubrication cannot be used, such as in situations of high contact pressure, cryogenic temperatures and vacuum (DONNET; ERDEMIR, 2004; ERDEMIR, 2005). The incorporation of a solid lubricant dispersed in a metallic matrix is not something new in the field of materials. Such components have found use in household appliances and various small pieces of equipment, such as blenders, electric shavers, drills, printers, among others (PARUCKER et al., 2014).

Bushings with self-lubricating materials, mainly with a copper or bronze matrix, containing molybdenum disulphide or graphite powder as particles, have been produced and used for decades in various engineering applications. However, these parts do not have high mechanical strength due to the high volumetric content (15% to 40%) of solid lubricant particles, which results in a low degree of continuity of the matrix phase, which is the structural element responsible for the material's mechanical strength. The high volumetric percentage of solid lubricant, which has low shear strength, does not contribute to the mechanical strength of the metal matrix (HAMMES, et al., 2014; BINDER, 2009; PARUCKER et al., 2014). In addition, the low hardness of the metal matrix allows solid lubricant particles to gradually clog up the contact surface of the material (HAMMES, 2011). As a result, these materials cannot be used for many typical mechanical applications, where greater mechanical and wear resistance of the self-lubricating material is required. In this sense, the development of sintered self-lubricating composites, which combine a low coefficient of friction with high mechanical strength, is becoming necessary for more specific engineering applications (HAMMES et al., 2014).

Among the possible ways to increase mechanical resistance are: formation of a liquid phase during sintering, in order to induce densification and rearrange the particles of the solid lubricant; addition of alloying elements in the composite matrix, in order to form more resistant phases or compounds; and, the production of composites by different compaction techniques, in order to produce components with a higher density (PARUCKER et al., 2014; DEMETRIO et al., 2015; STEINBACH et al., 2015; RAMOS FILHO et al., 2014; BINDER, 2009; HAMMES, 2011; BINDER et al., 2011; HAMMES et al., 2014; SCHROEDER et al., 2012; RIVERA, 2016; KLEIN et al., 2015; SCHROEDER et al., 2010; DE MELLO et al., 2013).

A study that contributed to the development of sintered self-lubricating materials was carried out by Hammes (HAMMES, 2011), who aimed to develop high-performance dry self-lubricating composites that simultaneously had a low coefficient of friction ($\mu \leq 0.2$) and high mechanical strength ($\sigma t \geq 250$ MPa). To do this, the author used two processing routes, double compaction (DC) and liquid phase sintering, as well as the addition of hardening alloying elements to the matrix (C, Ni and Mo) forming the Fe+1.5Mo+0.8C+4Ni+1Si alloy. The results are shown in Figure 3. For the first pair of samples (SC and DC) no solid lubricant was added, remaining the basic chemical composition (Fe+1.5Mo+0.8C+4Ni+1Si), thus justifying the higher friction coefficient results, for the other pairs of samples the following solid lubricants were added respectively: 5%hBN, 5% graphite and 5%hBN plus 5% graphite combined with 15%Cu to promote liquid phase sintering. The best results were achieved with the samples produced with the two solid lubricants combined with 15% Cu, which promoted liquid-phase sintering, and especially the 15Cu-5hBN/5C-DC sample which was processed using the double compaction technique, which, as has been seen in other studies (RECKNAGEL et al., 2011; GETHING et al., 2005), promotes greater densification and consequently increased mechanical strength (HAMMES, 2011).

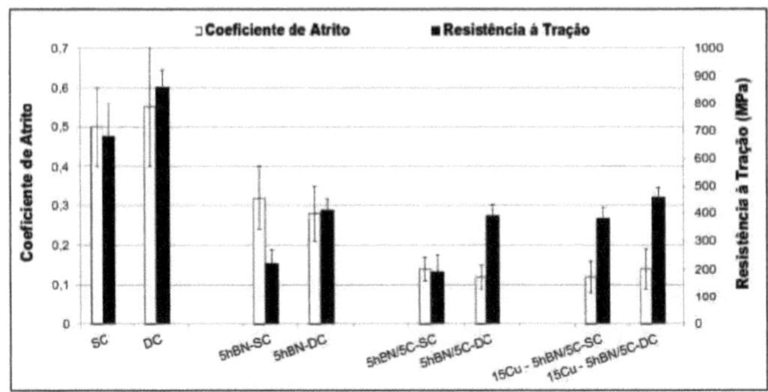

Figure 3 - Results of coefficient of friction and tensile strength of sintered composites produced by single compaction (SC) and double compaction (DC). All have a Fe+1.5Mo+0.8C+4Ni+1Si matrix (HAMMES, 2011).

The same author, in another paper, published a study on the effect of the compaction technique used (single compaction or double compaction) on the tribological behavior of a self-lubricating composite produced with hBN and graphite powder. The author concluded that: in constant sliding load tests, the coefficient of friction and wear rate are not affected by the compaction technique; and that double compaction increases durability, this effect being directly related to the reduction in porosity, which in turn causes an improvement in the mechanical resistance of self-lubricating composites (HAMMES et al., 2014).

Mucelin, et al (2014) studied the influence of a mixture of h-BN and graphite as solid lubricants on the tribological behavior of self-lubricating composites based on an Fe-0.5Si-C matrix produced by MP. The main results are shown in Figure 4. It can be seen that the addition of larger volumes of solid lubricant greatly improved durability, but the yield stress was affected. It was also found that decreasing the proportion of h-BN in the mixtures strengthened the tribological and mechanical properties (MUCELIN et al., 2014).

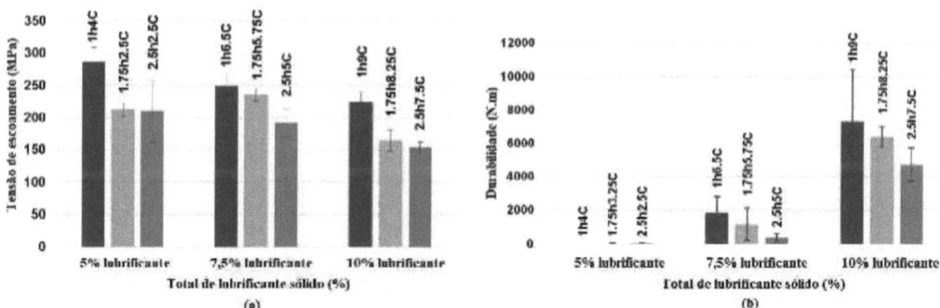

Figure 4 - Mechanical (a) and tribological (b) results of the samples studied (MUCELIN et al., 2014).

The use of hBN and graphite as a solid lubricant was also carried out by Ramos Filho and his colleagues in 2014. In this work, the effect of adding 0.8% phosphorus to a pure iron matrix containing different levels of graphite and hBN as solid lubricants (2.5 and 5.0% by volume) was investigated using dilatometry, micro-graphic analysis and tensile testing. The results showed that the addition of graphite reacts with the matrix to form pearlite, thus failing to form a composite with a solid lubricant (Figure 5(b)). The samples containing hBN showed stabilization of the alpha phase (Figure 5 (c)) and shrinkage of approximately 2% of the samples, which confirms the presence of densification-inducing mechanisms. With regard to the tensile results, only the sample with 2.5% hBN showed mechanical properties at reasonable levels (Figure 6) (RAMOS FILHO et al., 2014).

Figure 5 - Microstructure of the samples after sintering at 1125°C for 60 min. (a) Fe-0.8%P, (b) Fe-0.8%P-5.0% vol. graphite, (c) Fe-0.8%P-5% vol. hBN (RAMOS FILHO, et al., 2014).

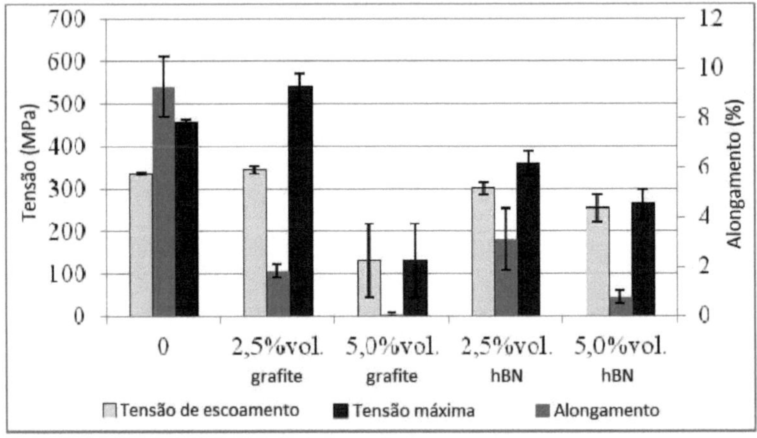

Figure 6 - Mechanical properties of Fe-0.8%P samples containing different amounts of graphite or hBN (RAMOS FILHO, et al., 2014).

Parucker, et al (2014) studied the stability and chemical interaction between a nickel matrix and three solid lubricant additives (MoS_2, graphite and hBN). The results showed thermodynamic stability during sintering for the compounds containing graphite and hexagonal boron nitride. However, the same was not observed for molybdenum disulphide (MoS_2), which reacted with the nickel matrix to form a new nickel disulphide phase (Ni_3S_2) allowing molybdenum to enter into solution with the matrix, suggesting that it should not be used as a solid lubricant in sintering at 1150 °C (PARUCKER et al., 2014).

In the same line of work, the reaction between the solid lubricant MoS2 and iron-based matrices during sintering was studied, as well as its influence on the properties of composites. In order to evaluate the influence of MoS2 on the iron matrix, various MoS2 contents were used (0.6, 1.8, 3.1, 4.4 and 5.7% by volume). After this, the MoS2 content was fixed at 5.7% and alloying elements (Cr, Ni, Mo, Si, P, Cu and C) were added in order to assess their influence on the matrix. The particle size of MoS2 was also evaluated. The investigations led to the following results: 1) for a pure iron matrix, the reaction of the solid lubricant MoS2 with the matrix already occurs at 850°C and the reaction products are iron sulphides or mixed iron-molybdenum sulphides, with the resulting Mo diffusing into the matrix; 2) for temperatures equal to or greater than 1050°C, iron sulphide constitutes a liquid phase in situ; 3) the initial concentration of MoS2 had an influence on the type of sulphide formed and the amount of liquid phase formed after the reaction, which affected the morphology of the pores and sulphides, but had no influence on the temperature at which the reaction between the lubricant and the matrix took place; 4) changing the particle size of MoS2 had no effect on the reaction mechanism, temperature or products formed; 5) adding phosphorus in the form of a pre-alloy seemed to affect the wettability of the liquid phase formed and prevent it from spreading; 6) the addition of alloying elements was unable to prevent the reaction of M0S2 with the matrix at the sintering temperature used (1150 °C) or contribute to the self-lubricating nature of the composite (FURLAN et al., 2015).

The search for components with a low coefficient of friction has not only been limited to the use of solid lubricants added to the matrix, but also to the incorporation of materials capable of dissociating during sintering and thus reducing the material's coefficient of friction. An example is the patent filed by (BINDER et al., 2011) in which one of the objectives was to obtain self-lubricating sintered steel products from carbides or carbonates, which are capable of generating, after dissociation during sintering, graphite nodules responsible for the material's low coefficient of friction (BINDER et al., 2011). The generation of solid lubricant particles in situ is shown schematically in Figure 7.

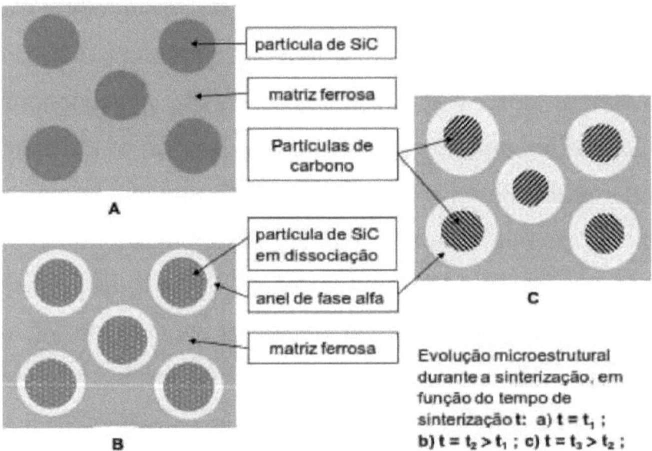

Figure 7 - Evolution of SiC particle dissociation and graphite particle formation during sintering (BINDER, 2009).

Figure 8 shows some of the results obtained by Binder (2009) in the production of self-lubricating composites with the formation of graphite in situ due to the addition of 3% SiC. In this work, materials were obtained with a friction coefficient of 0.04 for the Fe+0.6C+4Ni+1Mo+3SiC alloy, a wear rate of around $10x10^{(-6)}mm^3/Nm$ (of the material and the back body) and a maximum tensile strength of 1,000 MPa (BINDER, 2009).

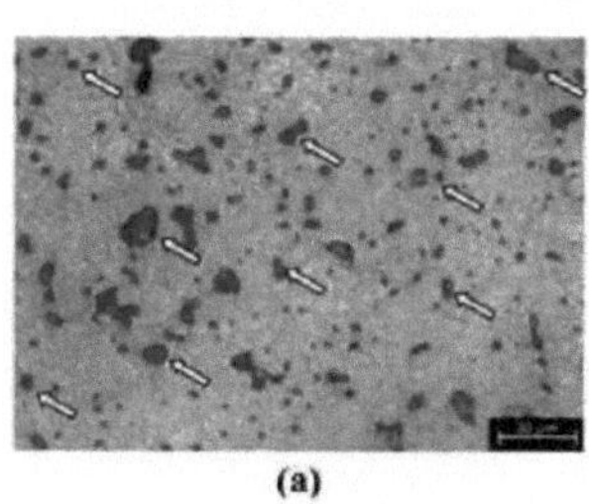

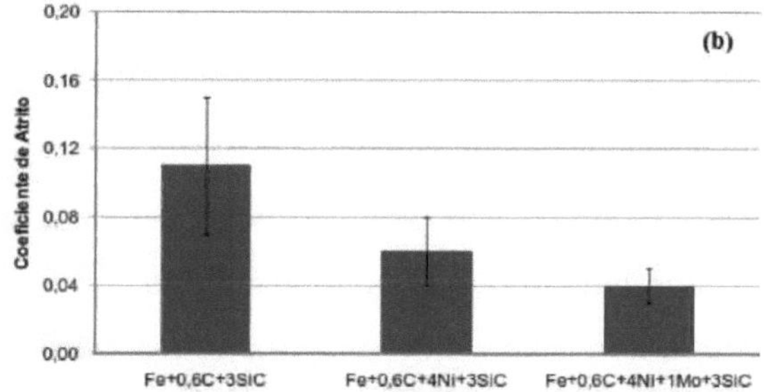

Figure 8 - (a) Typical microstructure of the self-lubricating alloy Fe-0.6C-4Ni-1Mo-3SiC (arrows indicate graphite nodules) (b) Coefficient of friction (BINDER, 2009)

As seen above, there are several possible ways to increase the mechanical strength of sintered self-lubricating composites, such as: promoting the formation of a liquid phase during sintering, in order to induce densification and rearrangement of the solid lubricant particles, and the production of composites by different compaction techniques, in order to produce components with a higher density. In addition, the introduction of alloying elements into the matrix of sintered self-lubricating composites can also produce a series of positive effects on the properties of the parts produced, due to the formation of more resistant phases or compounds. The controlled addition of alloying elements to the main component of the mixture is one of the most widely used means of achieving the desired mechanical quality. Most of the time, these elements are added during the mixing of the powders, through the use of alloy-bearing powders, and remain throughout the volume of the composite (PAVANATI, 2005).

However, in the processing of components by MPF, more than just the expected metallurgical effects on the properties of the component, the alloying elements exert a strong influence on the rate of contraction, powder compressibility and homogenization during sintering (GERMAN, 1996). In this sense, the use of some alloying elements in MPF presents various difficulties; molybdenum, for example, when used in its elemental state, presents a high sintering temperature and longer processing times to achieve a satisfactory degree of homogenization, while when added in the form of pre-alloyed powders it presents low compressibility (PAVANATI et al., 2007).

For some specific applications, such as hermetic compressors and automotive components, parts with wear in the order of a few micrometers (5-10 μm) already lose their engineering function and, in this sense, a modification of the chemical composition only on the surface of the material would already be necessary. Thus, modifications to the chemical composition on the surface of components produced by MPF have been studied and obtained using techniques based on plasma technology (JIANG et al., 2003; PAVANATI et al., 2007; BENDO et al., 2011; KLEIN et al., 2013; BENDO et al., 2014; BENDO et al., 2016).

2.4 PLASMA

DC plasma is considered to be an ionized gas made up of neutral and electrically charged species (0.1 to 1% of the total) such as electrons, ions, atoms and molecules in such a proportion that it remains electrically neutral (CHAPMAN, 1980).

Plasma arises when a potential difference is applied between two electrodes (cathode and anode) present in a hermetically sealed gaseous medium at low pressures. When a potential difference (DDP) is applied between these two electrodes, the ions and electrons present are accelerated by the electric field, colliding with other species and promoting their ionization, which in turn repeats the process, generating a cascade effect that enables the ionization of the system (LIEBERMAN; LICHTENBERG, 2005). Because of this production of charges, an electric current is generated which oscillates according to the DDP, causing the formation of various types of discharge, including the abnormal glow discharge which is the most suitable for generating plasma for the treatment and processing of materials. The three distinct regions that form the discharge are: the cathode sheath, the glow region and the anode sheath.

The cathode sheath is characterized by the strong electric field formed by the difference between the plasma potential and that applied to the cathode. This strong electric field allows the species formed in the luminescent region to be accelerated on their way to the cathode and thus collide with it (LIEBERMAN; LICHTENBERG, 2005). These collisions with the surface of the cathode can cause various effects, as can be seen in the diagram in Figure 9, among which the most important one for surface enrichment that should be commented on is *sputtering*. *Sputtering*, also known as sputtering, is the ejection of atoms from the cathode due to the repeated collisions between atoms caused by ionic impact. This phenomenon is extremely important for this work because it is from it that the surface enrichment technique is carried out (KLEIN et al., 2013).

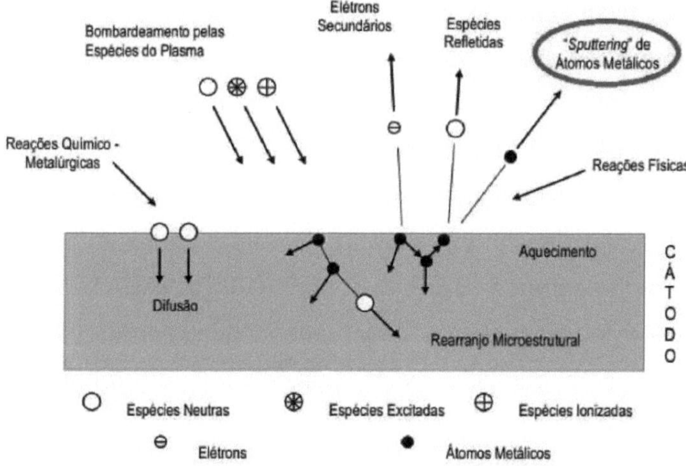

Figure 9 - Schematic of reactions near the cathode. Adapted from (SCHEUER, 2011).

The main function of surface enrichment is to add alloying elements to the surface of the material and thus promote modifications to its mechanical and tribological properties; modifications which are essential for optimizing self-lubricating composites. In this sense, the next topic presents how the surface enrichment technique is carried out.

2.4 SURFACE ENRICHMENT

In this process, the sample, inside a plasma reactor, is positioned in a confined anode-cathode configuration, i.e. it is positioned on a grounded anode and is surrounded by a cathode made of a suitable material. Inside the reactor, the cathode is bombarded by the energetic species of the plasma , producing heat, which in turn is transmitted to the sample mainly by radiation, thus promoting the activation of sintering mechanisms such as: evaporation; condensation; and the formation of a greater density of vacancies, thus activating self-diffusion in the volume close to the surface of the component (KLEIN, et al., 2013).

The bombardment of the energetic species in the plasma also causes the atoms in the cathode to be pulverized, thus promoting surface enrichment of the sample with the chemical elements in it. In this mechanism, the atoms of the cathode pass through the space between the cathode and the sample, depositing on the surface of the sample and diffusing into the sample (KLEIN et al., 2013). Figure 10 illustrates this mechanism. Thus, by bombarding the energetic species, it is possible to simultaneously sinter the components and enrich the surface of the sample with the alloying element determined by the chemical composition of the cathode (BENDO et al., 2014; PAVANATI et al., 2005). In this way, the alloying elements are only present on the surface, which is the region where the component is mainly subjected to environmental conditions and relative sliding.

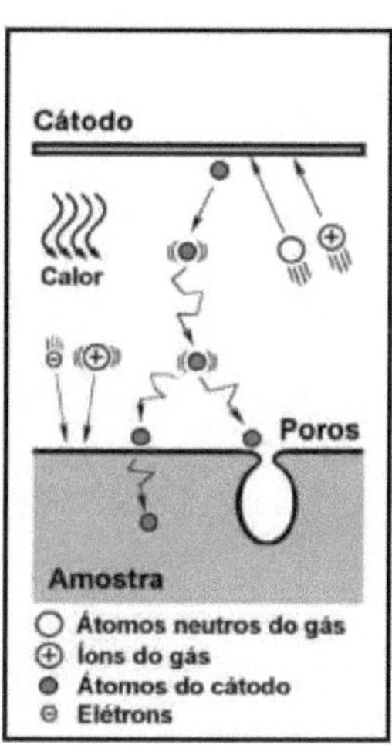

Figure 10 - Schematic drawing of the surface enrichment mechanism (PAVANATI et al., 2007).

The surface enrichment of sintered components has been studied and developed by several researchers (PAVANATI et al., 2005; PAVANATI et al., 2007; BENDO et al., 2014; MARCHIORI et al., 2007). In order to enlighten the reader about the work carried out in this area, a bibliographical review of authors who have covered this topic is presented below.

Pavanati, in his work, studied the surface enrichment of chromium in pure iron samples produced by PM during sintering using abnormal glow discharge in the confined anode-cathode configuration. The methods used were ABNT 430 stainless steel and ABNT 1020 carbon steel, the latter for comparative purposes. The results showed that the sample enriched with the ABNT 430 method had greater surface sealing (Figure 11 (a) and (b)) due to the formation of a chromium-enriched layer 15-20 µm deep, and the percentage of chromium reached on the surface of the samples was approximately 10% due to the stabilization of the α phase in them (Figure 11 (c)). The author also analyzed the effect of some processing parameters in relation to the thickness of the chromium layer formed and found that its thickness increased with the following parameters: a) enrichment time; b) applied voltage; c) enrichment temperature;

and d) pulse on time. In addition, it was observed that the percentage of chromium on the surface did not vary with changes in the processing parameters due to the stabilization of the α phase (PAVANATI et al., 2005).

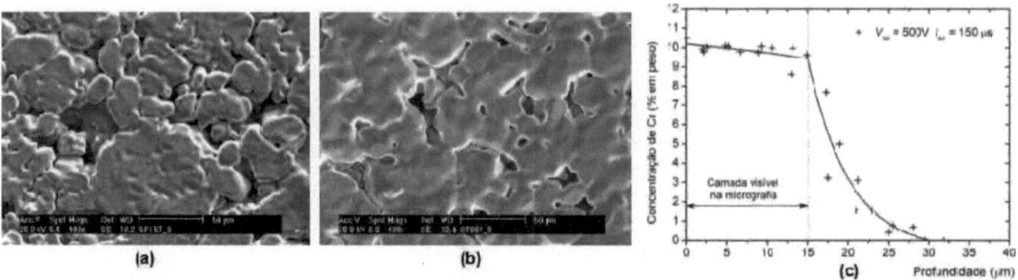

Figure 11 - Micrographs of the surface of the samples sintered using the ABNT 1020 carbon steel method (a) and the ABNT 430 stainless steel method (b) and the chromium concentration profile for the samples plasma sintered using the ABNT 430 stainless steel method (c) (PAVANATI et al., 2005).

In another paper, the same author studied the surface enrichment of chromium and molybdenum in sintered iron samples with different carbon contents (0.2, 0.4 and 1.0%C). Microstructural analysis showed that carbides formed on the surface both when enriched with chromium and molybdenum and that the layer of carbides formed became more evident the higher the carbon content present in the alloy composition (Figure 12). In the case of chromium enrichment, it was found that Cr7C3 carbides formed, but they formed discontinuously, producing a morphology containing cavities and microporosities. In the case of the molybdenum-enriched samples, the formation of mixed Fe3Mo3C carbides was observed. These carbides appeared in the form of acicular precipitates in the case of the sample containing 0.2%C and in the form of crystals nucleated on the surface for the samples with 0.45 and 1.0% carbon. The nucleation of these crystals was also responsible for the formation of the surface morphology with microcavities present in the regions where the grain contours were located (PAVANATI, 2006).

Figure 12 - Cross-sectional morphology of sintered samples enriched with Mo. (a) Pure iron, (b) Fe+0.2%C, (c) Fe+0.45%C and (d) Fe+1.0%C (PAVANATI, 2006).

The following year Pavanati published an article in which he discussed the stabilization of ferrite induced by the surface enrichment of Mo in sintered pure iron (PAVANATI et al., 2007). Pavanati proposed in Figure 13 how the Mo diffusion process evolves in an iron matrix during the sintering cycle. In (1) the sample is at room temperature and has an α phase throughout the entire volume. In (2) the α-γ phase transformation occurs due to heating (910 °C). In (3) due to the increase in temperature to 1150°C, Mo deposition becomes important and a slight enrichment of Mo occurs. However, it is assumed that this does not reach 4% by mass in the Fe matrix. In (4) the Mo enriched layer grows as well as its concentration on the surface, reaching around 4% by mass, stabilizing the α phase of the iron ("a" grains). In (5) the α-phase grains grow with a columnar morphology until they find a place that characterizes the grain boundary. In (6) the discharge has been switched off, the heating source and Mo atoms have been extinguished, but due to the presence of Mo in solid solution Fe-α phases nucleate in the pre-existing ferrite grains, but with a lower Mo concentration ("b" grains). While Mo is present in solid solution, these grains follow a columnar growth, similar to the pre-existing "a" grains. Finally, in (7) the ferrite grains follow a columnar growth until a certain depth where there is enough free energy to nucleate new grains (PAVANATI et al., 2007).

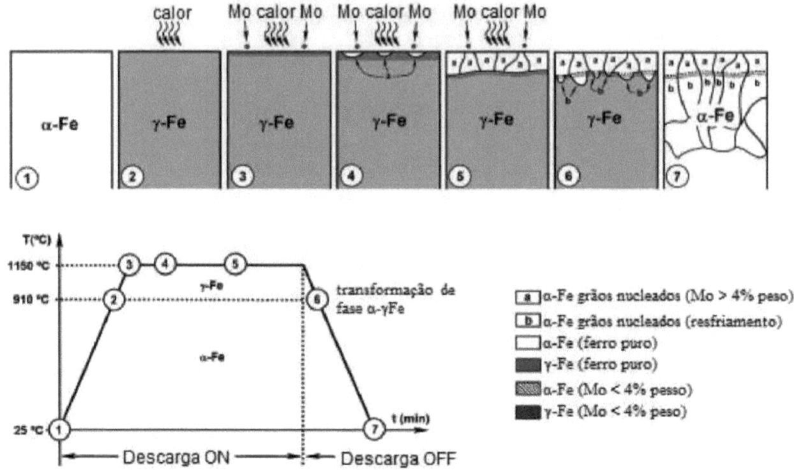

Figure 13 - Evolution of phase transformations during plasma sintering (PAVANATI et al., 2007).

In addition, Pavanati et al. (2007) found that as the tension applied to the method increased, there was an increase in the depth of the enriched layer (from 25µm to 75µm) and also in the Mo content present in the layer (Figure 14). According to him, these effects are directly related to the higher sputtering rates when the cathode voltage is increased, leading to a higher concentration of Mo atoms in the gas phase and, as a consequence, greater Mo deposition on the Fe samples (PAVANATI et al., 2007).

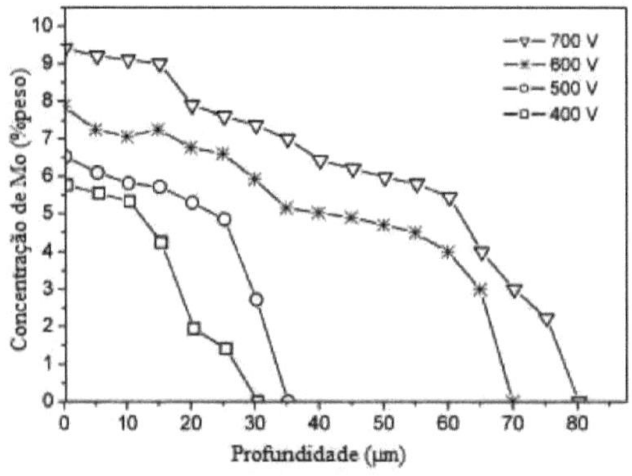

Figura 14 - Molybdenum concentration profile for different cathode voltages (PAVANATI et al., 2007).

Bendo also studied the enrichment of Mo in sintered pure iron. The results showed that on the surface of the layer the Mo concentration was in the order of 3.5% by mass, and remained at around 3.0 to 3.5% until a depth of 15 µm (Figure 15). From this point there was a sharp drop in molybdenum content up to a depth of 25 µm. This effect has already been observed by (PAVANATI et al., 2007) and is related to the stabilization of the iron alpha phase at the sintering temperature (BENDO, 2009).

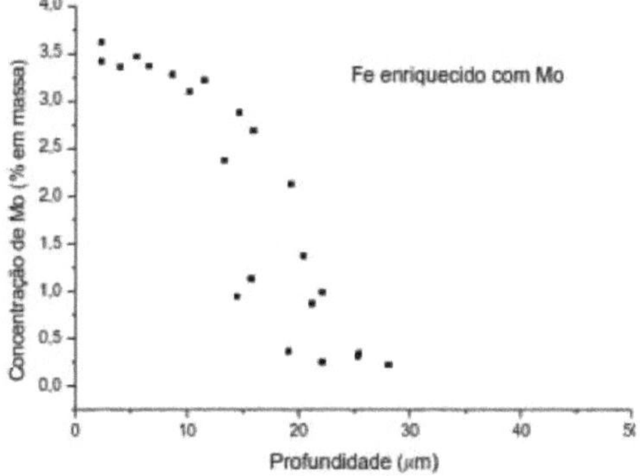

Figura 15 - Average molybdenum concentration profile of sintered and plasma-enriched iron samples
iron samples (BENDO, 2009).

Hammes (2006) developed systematic studies of the process of enriching surfaces with molybdenum in pure iron samples during plasma sintering. The author evaluated the effects of treatment temperature and plasma potential configuration on Mo deposition (cathode, anode or floating potential). With regard to the effect of the electrode configuration, it was found that, contrary to what was imagined, the fact that the sample was on the cathode did not represent an increase in the Mo deposition rate when compared to the other two configurations (anode and floating potential). With regard to the effect of temperature, it was observed that the lower the treatment temperature, the greater the amount of Mo on the surface and the smaller the size of the agglomerates formed (Figure 16). According to the author, this greater amount of Mo on the surface is due to the low diffusion of Mo at 800ºC, which then accumulates on the surface. Similarly, it was observed that the higher the enrichment temperature, the thicker the layer due to the greater diffusion of the element (HAMMES, 2006). This is illustrated in Figure 17. It was also noticeable that below the Mo concentration at which stabilization of the α phase occurs (around 3%), the Mo profile fell sharply, because in the γ phase the diffusion of the element is significantly lower as commented by (PAVANATI et al., 2007).

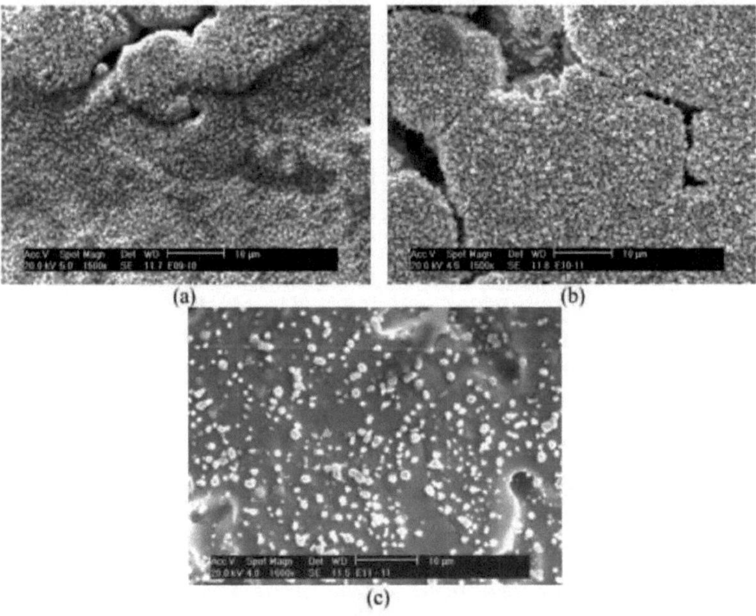

Figura 16 - Micrographs of the surfaces of 700 V anode-enriched samples. (a) 800°C, (b) 1000 °C and (c) 1150 °C (HAMMES, 2006).

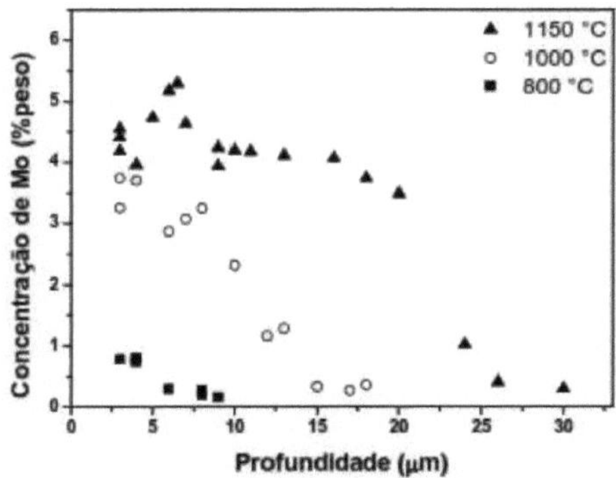

Figura 17 - Mo concentration profiles of samples treated in the confined anode-cathode configuration (HAMMES, 2006).
confined cathode configuration (HAMMES, 2006).

Still on the subject of Hammes' work (2006), the author evaluated the effect of bombardment energy on the Mo concentration profiles of samples enriched at 500 and 700 V, at a temperature of 1150 °C. The results showed that a higher deposition rate and greater penetration of Mo is achieved in the material when the discharge voltage is increased, since the sputtering is more intense and therefore there is a greater chemical potential for molybdenum in the treatment atmosphere (Figure 18 (a)). Hammes also evaluated that with the increase in pressure from 1 Torr to 3.5 Torr there was a greater deposition of Mo, as well as a greater enriched layer (Figure 18 (b)). According to the author, it is assumed that with the increase in gas pressure in the chamber, there was a greater number of ions bombarding the cathode, which favored sputtering (HAMMES, 2006).

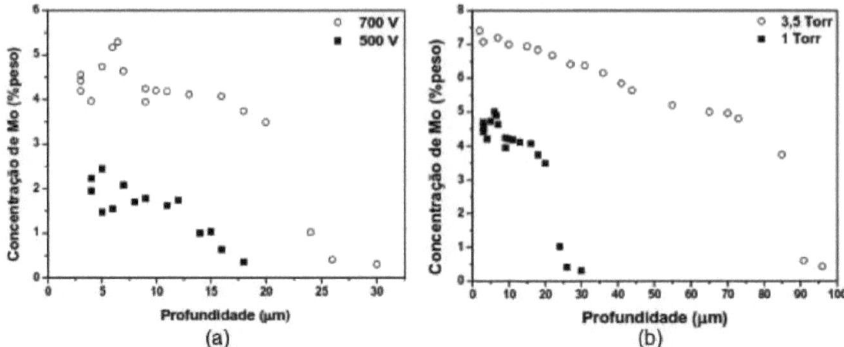

Figure 18 - Mo concentration profile of samples enriched at 1150 °C (a) at 500 V and 700 V and 1 Torr and (b) 1 Torr and 3.5 Torr at 700V (HAMMES, 2006).

Lawall also studied the sintering process of pure iron compacts in electric discharge in the confined anode-cathode configuration. One of the variables studied was the percentage of argon in the gas mixture. The results showed that there was an increase in the concentration of chromium deposited as a result of the increase in the proportion of argon in the gas mixture, which is due to the greater efficiency of *sputtering*, since according to (CHAPMAN, 1980) the *sputtering* coefficient increases with the mass of the ion bombarding the cathode. The heavier the argon ion, then

than hydrogen, the concentration of pulverized atoms increases as the proportion of argon in the gas mixture increases (LAWALL, 2001).

The study of Ni enrichment was carried out by Cardoso (2003) and Marchiori (2007). Cardoso carried out an experimental and theoretical study of Ni surface enrichment on IF steel and sintered iron. The numerical and experimental results showed good agreement with each other (Figure 19). X-ray diffraction analysis showed the presence of taenite on the surface of the samples. In addition, for the sintered samples, a small closure of surface pores was observed due to the deposition of nickel on the edges of the pores. This effect was clearly seen in cases where the deposition rate was higher (700V) (CARDOSO, 2003).

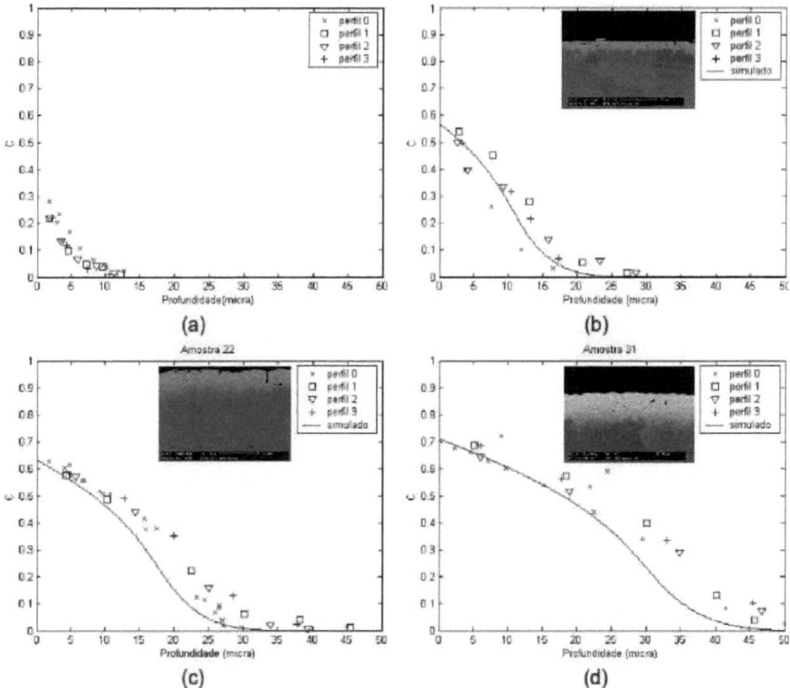

Figure 19 - Concentration profiles of the treated samples and their micrographs. 1150°C, 700V, 1.95Torr, 20%H_2/80%Ar, 180 µs, 20 sccm (a) 0 min, (b) 30 min, (c) 60 min, (d) 120 min.

Marchiori studied the corrosion resistance of pure iron samples sintered and enriched with Ni. The samples enriched in anode showed the same results as those obtained by (CARDOSO, 2003), i.e. a concentration of Ni on the surface of around 60% (Figure 20) and the formation of taenite phases (NiFe) (Figure 21). In addition, during sintering, the diffusion of Ni into the matrix resulted in the formation of a Ni enriched layer. However, no increase in corrosion resistance was observed due to the formation of this nickel-rich layer. According to the author, this is possibly because the Ni rich phases are discontinuous or because the corrosion test was inadequate for the situation studied (MARCHIORI et al., 2007).

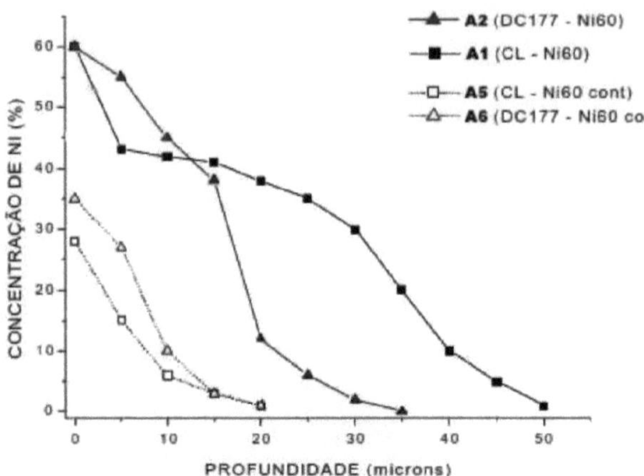

Sample 1: CL powder: anode configuration for 00 minutes;

Sample 2: DC177 powder; anode configuration for 60 minutes;

Sample 5: CL powder; anode configuration (00 min), with contaminated Ni method;

Sample 6: DC177 powder; anode configuration (00 min), with containerized Ni method

Figura 20 - Ni concentration profiles in the samples (MARCHIORI et al., 2007).

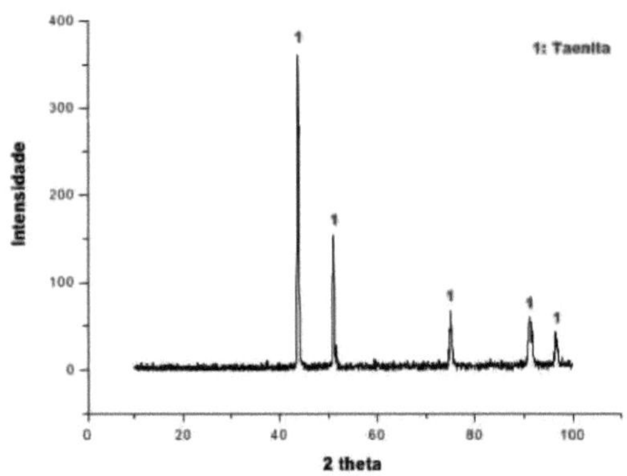

Figure 21 - X-ray diffractometry of the Ni enriched surface in the anode-cathode configuration confined for 60 minutes (MARCHIORI et al., 2007).

Based on what has been said so far, it can be seen that modifications to the composition and morphology of the surfaces of parts produced by ferrous powder metallurgy (FPM) have been achieved through the use of surface enrichment (HAMMES et al., 2006; CARDOSO, 2003; PAVANATI et al., 2007; BENDO et al., 2011; BENDO et al., 2014; BENDO et al., 2016). The surface enrichment, as already mentioned, occurs with the addition of alloying elements to the surface of the sintered component and can be carried out simultaneously with the sintering stage, so there are no additional processing costs, but only the production of the cathode that can be used several times.

In this context, surface enrichment has emerged as an attractive process for increasing mechanical and tribological resistance in self-lubricating components, because in this process the addition of alloying elements occurs only in the region where it is actually needed, i.e. on the surface and just below it. In addition, as the process is carried out simultaneously with the sintering stage, there is no increase in processing time and a reduction in costs in relation to the raw material used, since it is not necessary to use powders to insert the alloying elements into the material.

3 FINAL CONSIDERATIONS

Based on what has been discussed so far, the enrichment of self-lubricating composites has some advantages, such as:

- Increasing the strength of the self-lubricating composite only in the area of interest (surface and subsurface regions).
- Enrichment takes place during a stage that was already necessary (sintering), so there are no external costs due to the enrichment stage, and the only cost that is necessary is for the method, which can be used repeatedly.
- There are cost savings due to not using connected or pre-connected powders.
- Not using bonded or pre-bonded powders also facilitates the processing of the self-lubricating composite because it avoids reducing the compressibility of the powder and problems with the contraction rate and homogenization of the component.

In this sense and based on the above, plasma surface enrichment appears as a potential technique for optimizing self-lubricating components, thus launching new material solutions for use in various engineering sectors related to wear problems and mechanical and energy losses (friction).

BIBLIOGRAPHICAL REFERENCES

BENDO, T. **Plasma nitriding of pure iron surface enriched with molybdenum.** 2009. 65 p. Dissertation (Master's Degree) - Federal University of Santa Catarina, Technology Center. Postgraduate Program in Materials Science and Engineering, Florianópolis, 2009.

BENDO, T. et al. Plasma Nitriding of Surface Mo-Enriched Sintered Iron. **Isrn Materials Science,** [s.l.], v. 2011, p.1-8, 2011. Hindawi Publishing Corporation. http://dx.doi.org/10.5402/2011/121464.

BENDO, T. et al. Nitriding of surface Mo-enriched sintered iron: Structure and morphology of compound layer. **Surface And Coatings Technology,** [s.l.], v. 258, p.368-373, nov. 2014. Elsevier BV.

http://dx.doi.org/10.1016Zj.surfcoat.2014.08.072.

BENDO, T. et al. The effect of Mo on the characteristics of a plasma nitrided layer of sintered iron. **Applied Surface Science,**[s.l.], v. 363, p.29-36, feb. 2016. Elsevier BV. http://dx.doi.org/10.1016/j.apsusc.2015.11.200.

BINDER, C. **Development of new types of self-lubricating dry sintered steels with high mechanical strength and low coefficient of friction via powder injection moulding.** 170 f. Thesis (PhD) - Federal University of Santa Catarina, Technology Center, Postgraduate Program in Materials Science and Engineering, Florianópolis, 2009

BINDER, Cristiano et al. 'Fine tuned' steels point the way to a focused future. **Metal Powder Report,** [s.l.], v. 65, n. 4, p.29-37, May 2010. Elsevier BV. http://dx.doi.org/10.1016/s0026-0657(10)70108-9.

BINDER, R. et al. **Composition of particulate materials for forming selflubricating products in sintered steels, product in self- lubricating sintered steel and process for obtaining self-lubricating products in**

sintered steel. US n° 20110286873A1, Dec. 9, 2009, Nov. 24, 2011. United States Patent and Trademark Office, 2011.

Binder, R. et al. **Metallurgical composition of particulate materials, self-lubricating sintered product and process for obtaining**

self-lubricating sintered products. BR PI n° 0803956-9, September 12, 2008, September 22, 2010 INPI, 2008.

CARDOSO, Rodrigo Perito. **Numerical and experimental study of the deposition and diffusion process of nickel via plasma in iron samples in the confined anode-cathode configuration.** Florianópolis, 2003. 104 p. Dissertation (Master's Degree) - Federal University of Santa Catarina, Technology Center. Postgraduate Program in Mechanical Engineering, Florianópolis, 2003.

CHAPMAN, B. **Glow Discharge Processes:** Sputtering and Plasma Etching. [S.l.]: John Wiley & Sons, 1980. 432 p. ISBN 978-0-471-07828-9.

DE MELLO, J. B.; BINDER, R. A methodology to determine surface durability in multifunctional coatings applied to soft substrates. **Tribology International,** [s.l.], v. 39, n. 8, p.769-773, aug. 2006. Elsevier BV.

http://dx.doi.org/10.1016Zj.triboint.2005.07.015.

DE MELLO, J. D. B. et al. Effect of the metallic matrix on the sliding wear of plasma assisted debinded and sintered MIM self-lubricating steel. **Wear,** [s.l.], v. 301, n. 1-2, p.648-655, abr. 2013. Elsevier BV.

http://dx.doi.org/10.1016/j.wear.2013.01.011.

DE MELLO, J. D. B. et al. **Influence of sintering temperature on the tribological behavior of plasma assisted debinded and sintered MIM self lubricating steels**. ASME 2010 10th Biennial Conference on Engineering Systems Design and Analysis. Istanbul: [s.n.]. 2010. p. 373-380.

DE MELLO, J. D. B. et al. Effect of precursor content and sintering temperature on the scuffing resistance of sintered self lubricating steel. **Wear,** [s.l.], v. 271, n. 9-10, p.1862-1867, jul. 2011. Elsevier BV. http://dx.doi.org/10.1016/j.wear.2010.11.038.

DEMETRIO, V. B. et al. **Development of self-lubricating steels by compression of granulated powders**. PTECH 2015 - Tenth International Latin American Conference on Powder Technology. Mangaratiba: [s.n.]. 2015.

DENG, J.; CAN, T.; SUN, J.. Microstructure and mechanical properties of hot-pressed Al2O3/TiC ceramic composites with the additions of solid lubricants. **Ceramics International,** [s.l.], v. 31, n. 2, p.249-256, jan. 2005. Elsevier BV. http://dx.doi.org/10.1016/j.ceramint.2004.05.009.

DONNET, C; ERDEMIR, A. Historical developments and new trends in tribological and solid lubricant coatings. **Surface And Coatings Technology,** [s.l.], v. 180-181, p.76-84, mar. 2004. Elsevier BV. http://dx.doi.org/10.1016Zj.surfcoat.2003.10.022.

DONNET, C.; ERDEMIR, A.. Solid Lubricant Coatings: Recent Developments and Future Trends. **Tribology Letters,** [s.l.], v. 17, n. 3, p.389-397, oct. 2004. Springer Nature. http://dx.doi.org/10.1023/b:tril.0000044487.32514.1d.

ERDEMIR, A. Solid lubricants and self lubricating films. In: BHUSHAN, B. **Modern Tribology Handbook**. [S.l.]: CRC Press, v. II, 2001. p. 787-825.

ERDEMIR, A. Review of engineered tribological interfaces for improved boundary lubrication. **Tribology International,** [s.l.], v. 38, n. 3, p.249-256, mar. 2005. Elsevier BV. http://dx.doi.org/10.1016/j.triboint.2004.08.008.

ERDEMIR, A.; ERYILMAZ, O. L.; KIM, S. H.. Effect of tribochemistry on lubricity of DLC films in hydrogen. **Surface And Coatings Technology,** [s.l.], v. 257, p.241-246, Oct. 2014. Elsevier BV.

http://dx.doi.org/10.1016/j.surfcoat.2014.08.002.

FURLAN, K. P. et al. **Process for obtaining a ferrous matrix self-lubricating component and a self-lubricating component**. BR n° 1020140248404, 10 Oct. 2014, INPI, 2014.

FURLAN, K. P. et al. Influence of alloying elements on the sintering thermodynamics, microstructure and properties of Fe-MoS2 composites. **Journal Of Alloys And Compounds,** [s.l.], v. 652, p.450-458, dec. 2015. Elsevier BV. http://dx.doi.org/10.1016/jjallcom.2015.08.242.

GERMAN, R. M. **Sintering Theory and Practice**. New York: John Wiley & Sons, 1996. 568 p. ISBN 978-0-471-05786-4.

GETHING, B. A. et al. The effect of nickel on the mechanical behavior of molybdenum P/M steels. **Materials Science And Engineering: A,** [s.l.], v. 390, n. 1-2, p.19-26, jan. 2005. Elsevier BV.

http://dx.doi.org/10.1016/j.msea.2004.05.087.

GOGOTSI, Y. et al. Conversion of silicon carbide to crystalline diamond-structured carbon at ambient pressure. **Nature,**[s.l.], v. 411, n. 6835, p.283287, May 17, 2001. Springer Nature. http://dx.doi.org/10.1038/35077031.

HAMMES, G. et al. Effect of double pressing/double sintering on the sliding wear of self-lubricating sintered composites. **Tribology International,** [s.l.], v. 70, p.119-127, feb. 2014. Elsevier BV. http://dx.doi.org/10.1016/j.triboint.2013.09.016.

HAMMES, G. **Modification of the Chemical Composition of the Surface of Metallic Components via Plasma Catodic Spraying: Equipment Design and Process Development.** 2006. 71 f. Dissertation (Master's Degree) - Materials Engineering Course, Department of Engineering Mecânica, Universidade Federal de Santa Catarina, Florianópolis, 2006.

HAMMES, G. **S Development of self-lubricating dry sintered composites with granulated solid lubricants**. 2013. 25 p. Report (Post-doctorate) - Materials Engineering Course, Department of Mechanical Engineering, Universidade Federal de Santa Catarina, Florianópolis, 2013.

HAMMES, G. **Self-lubricating dry sintered steels with high mechanical strength associated with low coefficient of friction**. 2011. 89 f. Thesis (PhD) - Materials Engineering Course, Department of Mechanical Engineering, Universidade Federal de Santa Catarina, Florianópolis, 2011.

HAMMES, G. et al. **Plasma Sintering of Pure Iron with Simultaneous Surface Enrichment of Mo with Samples Positioned at the Cathode, Anode, or Floating Potential.** 61st ABM ANNUAL CONGRESS. Rio de Janeiro: [s.n.]. 2006.

HAMMES, G. et al. Fe-hBN Composites Produced by Double Pressing and Double Sintering. **Materials Science Forum,**[s.l.], v. 802, p.311-316, Dec. 2014. Trans Tech Publications.

http://dx. doi. org/10.4028/www. scientific. net/msf. 802.311.

HOGMARK, S. et al. Mechanical and tribological requirements and evaluation of coating compositesB. Bhushan. In: BHUSHAN, B. **Modern Tribology Handbook**. [S.l.]: CRC Press, v. II, 2001. p. 931-959.

HOLMBERG, K.; ANDERSSON, P.; ERDEMIR, A. Global energy consumption due to friction in passenger cars. **Tribology International,** [s.l.], v. 47, p.221-234, mar. 2012. Elsevier BV. http://dx.doi.org/10.1016/j.triboint.2011.11.022.

JIANG, X. et al. Multi-element Ni-Cr-Mo-Cu surface alloyed layer on steel using a double glow plasma process. **Surface And Coatings Technology,** [s.l.], v. 168, n. 2-3, p.142-147, May 2003. Elsevier BV. http://dx.doi.org/10.1016/s0257-8972(03)00008-

JOST, H. P. **Lubrication (tribology) : education and research; a report on the present position and industry's needs.** Department of Education and Science. London, p. 79. 1966.

JOST, H. P. Tribology - Origin and future. **Wear,** [s.l.], v. 136, n. 1, p.1-17, feb. 1990. Elsevier BV. http://dx.doi.org/10.1016/0043-1648(90)90068-I.

KLEIN, A. N. et al. Economic Opportunities for Advanced Materials in Tribology. In: ASSUNÇÂO, F. C. R. **Advanced Materials in Brazil** 20102022. Brasilia: Center for Strategic Studies and Management, 2010. ISBN 97885-60755-25-7.

KLEIN, A. N. et al. Thermodynamic aspects during the processing of sintered materials. **Powder Technology,** [s.l.], v. 271, p.193-203, feb. 2015. Elsevier BV. http://dx.doi.org/10.1016Zj.powtec.2014.11.022.

KLEIN, A. N. et al. DC Plasma Technology Applied to Powder Metallurgy: an Overview. **Plasma Science And Technology,**[s.l.], v. 15, n. 1, p.70-81, jan. 2013. IOP Publishing. http://dx.doi.org/10.1088/1009-0630/15/1/12.

KORENBLIT, Y.; YUSHIN, G. Carbide-Derived Carbons. In: GOGOTSI, Y.; PRESSER, V. **Carbon Nanomaterials**. [S.l.]: CRC Press, v. II, 2014. p. 529. ISBN 9781466502420.

KUBOTA, M. **Report by the committee on tribology standardization**.

Association of Machinery Industry of Japan. Tokyo. 1982.

LAWALL, I. T. **Study of the sintering process of iron compacts in electrical discharge in the confined anode-cathode configuration**.

Florianópolis, 2001. xiv, 110 f. Thesis (Doctorate) - Federal University of Santa Catarina, Technology Center. Florianópolis, 2001

LIEBERMAN, M. A.; LICHTENBERG, A. L. **Principles of plasma discharges**

and. New Jersey: Jhon Wiley & Sons, 2005. 757 p.

MAHATHANABODEE, S. et al. Effects of hexagonal boron nitride and sintering temperature on mechanical and tribological properties of SS316L/h-

BN composites. **Materials & Design,** [s.l.], v. 46, p.588-597, Apr. 2013. Elsevier BV. http://dx.doi.Org/10.1016/j.matdes.2012.11.038

MARCHIORI, R. et al. Corrosion study of plasma sintered unalloyed iron: The influence of porosity sealing and Ni surface enrichment. **Materials Science And Engineering: A,** [s.l.], v. 467, n. 1-2, p.159-164, Oct. 2007. Elsevier BV. http://dx.doi.org/10.1016Zj.msea.2007.02.090.

MCNALLAN, M. et al. Nano-structured carbide-derived carbon films and their tribology. **Tinshhua Sci. Technol.,** [s.l.], v. 10, n. 6, p.699-703, dec. 2005. Institute of Electrical and Electronics Engineers (IEEE).

http://dx.doi.org/10.1016/s1007-0214(05)70138-3.

MOLGAARD, J. **Economic losses due to friction and wear-research and development strategies**. National Research Council Canada, Associate Committee on Tribology. [1984.

MUCELIN, K. J. et al. **Tribological study of self-lubricating composites with hexagonal boron nitride and graphite as solid lubricants**. TriboBR 2014 - Second International Brazilian Conference on Tribology. Foz do Iguaçu: [s.n.]. 2014.

PARUCKER, M. L. et al. Development of self-lubricating composite materials of nickel with molybdenum disulfide, graphite and hexagonal boron nitride processed by powder metallurgy: preliminary study. **Materials**

Research, [s.l.], v. 17, p.180-185, Aug. 2014. FapUNIFESP (SciELO). http://dx.doi.org/10.1590/s1516-14392013005000185.

PAULEAU, Y.; THIÈRY, F.. Deposition and characterization of nanostructured metal/carbon composite films. **Surface And Coatings Technology,** [s.l.], v. 180-181, p.313-322, mar. 2004. Elsevier BV. http://dx.doi.org/10.1016/j.surfcoat.2003.10.077.

PAVANATI, H. C. **Sintering of pure iron with simultaneous surface enrichment of chromium in abnormal electrical discharge.** 2005. 190 f. Thesis (Doctorate) - Materials Engineering Course, Department of Mechanical Engineering, Federal University of Santa Catarina, Florianópolis, 2005.

PAVANATI, H. C. **Sintering with surface enrichment of Cr or Mo by plasma in steel compacts with different carbon contents**. 2006. 28 p. Report (Post-doctorate) - Engineering Course of

Materials, Department of Mechanical Engineering, Federal University of Santa Catarina, Florianópolis, 2006.

PAVANATI, H. C. et al. Sintering unalloyed iron in abnormal glow discharge with superficial chromium enrichment. **Materials Science And Engineering: A,** [s.l.], v. 392, n. 1-2, p.313-319, feb. 2005. Elsevier BV.

http://dx.doi.org/10.1016Zj.msea.2004.09.034.

PAVANATI, H. C. et al. Ferrite stabilization induced by molybdenum enrichment on the surface of unalloyed iron sintered in an abnormal glow discharge. **Applied Surface Science,** [s.l.], v. 253, n. 23, p.9105-9111, set. 2007. Elsevier BV. http://dx.doi.org/10.1016/j.apsusc.2007.05.036.

PINKUS, O. et al. **Strategy for energy conservation through tribology**. New York: American Society of Mechanical Engineers, 1977. 174 p.

PRABHU, T. Ram. Effects of solid lubricants, load, and sliding speed on the tribological behavior of silica reinforced composites using design of experiments. **Materials & Design,** [s.l.], v. 77, p.149-160, jul. 2015. Elsevier

BV. http://dx.doi.org/10.1016Zj.matdes.2015.03.059.

RAMOS FILHO, A. I. et al. Development of Fe-P Matrix Composites Containing Embedded Solid Lubricants.**Materials Science Forum,** [s.l.], v. 802, p.96-101, dec. 2014. Trans Tech Publications.

http://dx.doi.org/10.4028/www.scientific.net/msf.802

RECKNAGEL, et al. **Higher Densities of PM-Steels by Warm Secondary Compaction and Sizing**. Euro PM2011 - Powder Pressing. [S.l.]: [s.n.]. 2011.

RIVERA, N. I. A. **Influence of heat treatments on the microstructural evolution and mechanical properties of self-lubricating sintered steel**. 2016. 90 p. Dissertation (Master's Degree) - Federal University of Santa Catarina, Technology Center, Postgraduate Program in Materials Science and Engineering, Florianópolis, 2016.

RODRIGUEZ, V. et al. Influence of solid lubricants on tribological properties of polyetheretherketone (PEEK). **Tribology International,** [s.l.], v. 103, p.4557, nov. 2016. Elsevier BV. http://dx.doi.org/10.1016/j.triboint.2016.06.037.

SCHEUER, C. J. **Plasma-assisted low-temperature cementation of AISI 420 martensitic stainless steel**. 2011. 92 p. Dissertation

(Master's Degree) - Federal University of Paranâ, Technology Sector. Postgraduate Program in Mechanical Engineering, Curitiba, 2011.

SCHROEDER, Renan et al. Internal lubricant as an alternative to coating steels. **Metal Powder Report,** [s.l.], v. 65, n. 7, p.24-31, nov. 2010. Elsevier BV. http://dx.doi.org/10.1016/s0026-0657(11)70043-1.

SCHROEDER, R. et al. Powder Injection Molding of Multimaterial Parts - Self Lubricating Steel Combined with Plain-Carbon Steel. **Materials Science Forum,** [s.l.], v. 727-728, p.243-247, Aug. 2012. Trans Tech Publications. http://dx.doi.org/10.4028/www.scientific.net/msf.727-728.243.

STEINBACH, R. et al. **Effect of liquid phase during sintering on mechanical and tribological properties of self-lubricating composites**. PTECH 2015 - Tenth International Latin American Conference on Powder Technology. Mangabira: [s.n.]. 2015.

THÜMMLER, F.; OBERACKER, R. **Introduction to Powder Metallurgy**. Lodon: Great Britain, v. 1, 1993.

WELZ, S.; GOGOTSI, Y.; MCNALLAN, M. J. Nucleation, growth, and graphitization of diamond nanocrystals during chlorination of

carbides. **Journal of Applied Physics**, [s.l.], v. 93, n. 7, p.4207-4217, 2003. AIP Publishing. http://dX.doi.org/10.1063/1.1558227.

ZALAZNIK, M. et al. Effect of the type, size and concentration of solid lubricants on the tribological properties of the polymer PEEK. **Wear,** [s.l.], v. 364-365, p.31-39, Oct. 2016. Elsevier BV.

http://dx.doi.org/10.1016Zj.wear.2016.06.013.

Printed by Books on Demand GmbH, Norderstedt / Germany